YOUR KNOWLEDGE HAS VALUE

- We will publish your bachelor's and master's thesis, essays and papers

- Your own eBook and book - sold worldwide in all relevant shops

- Earn money with each sale

Upload your text at www.GRIN.com
and publish for free

Bibliographic information published by the German National Library:

The German National Library lists this publication in the National Bibliography; detailed bibliographic data are available on the Internet at http://dnb.dnb.de .

Imprint:

Copyright © 2018 GRIN Verlag
Print and binding: Books on Demand GmbH, Norderstedt Germany
ISBN: 9783668818033

This book at GRIN:

https://www.grin.com/document/444909

Moritz Stüber

The Agroecology Movement in Costa Rica. Aims, Actors, Structures and Relation to Organic Agriculture

GRIN Verlag

Project in Organic Agriculture and Food Systems 2017/2018

The Agroecology Movement in Latin America

Aims, Actors, Structures and Relation to Organic Agriculture

Submitted by
Moritz Stüber

Table of content

List of tables

List of figures

List of abbreviations

ADM	Archer Daniels Midland company
CELAC	Community of Latin American and Caribbean States
CR	Costa Rica
e.g.	*exempli gratia* (for example)
etc.	*et cetera* (and so on)
EU	European Union
FAO(STAT)	Food and Agriculture Organization (Corporate Statistical Database)
GDP	Gross Domestic Product
GMO	Genetically Modified Organism
ha	hectare
HDI	Human Development Index
IFOAM	International Federation of Organic Agriculture Movements
Mio.	Million
MST	Landless Worker's Movement ("*Movimento dos Trabalhadores Sem Terra*")
OA	Organic Agriculture
OECD	Organisation for Economic Co-operation and Development
NGO	Non-Governmental Organisation
R&D	research and development
SAP	Structural Adjustment Programme
US(A)	United States (of America)

Glossary

Abroad

CAC	Farmer-to-farmer ("*Campesino-a-campesino*"), movement within Agroecology and commonly used method of knowledge exchange
CATIE	Tropical Agricultural Research and Higher Education Center cat("*Centro Agronómico Tropical de Investigación y Enseñanza*")
Kaqchikel	Indigenous Maya people in Guatemala
JICA	Japan International Cooperation Agency
MAELA	Agroecological Movement of Latin American and the Caribbean
PCaC	Farmer-to-Farmer (CAC) Programme, implemented by the UNAG
promotor	Agroecological trainer, to organise and give workshops
trueque	Barter, exchange of goods
UNAG	National Union of Farmers and Ranchers of Nicaragua,

Public policy sector, instruments (CR) and universities

ANAO	National Association of Organic Agriculture
Law No. 8408	Law on Sustainable Production (by MAG)
Law No. 8591	Law on Organic Agricultural Production (by MAG)
MAG	Ministry of Agriculture and Livestock
MINAE	Ministry of Environment, Energy and Telecommunications
PEN	State of the Nation Programme (*"Programa Estado de la Nación"*)
PNAO	National Programme of Organic Agirculture (by MAG)
PROCOMER	Export Promotion Agency of Costa Rica (*"Promotora del Comercio Exterior de Costa Rica"*)
RBA	Recognition of Environmental Benefits (implemented by MINAE)
RBAO	Recognition of Environmental Benefits on Organic Production (implemented by MAG)
UNA	National University of Costa Rica
UCR	University of Costa Rica

NGOs, research centres and private companies (CR)

ANAI	ANAI association, corporation for rural development (partner organisation of APPTA)
APPTA	Small-scale producer association of the Costa Rican Talamanca region (*"Asociacion de Pequeños Productores de Talamanca"*)
Cadena Agro-Alimentaria	Project "Agrifood Chain", developed by the CEDECO
CEDECO	Educative Corporation of Development in Costa Rica
CUDECA – ECOS	Consultancy Services for Development - Capacity, Organization and Sustainability Team
IBS Soluciones Verdes	Consulting Agency for social and environmental issues
MAOCO	Costa Rican Organic Agricultural Movement
Red COPROALDE	Costa Rican Network of Rural Organisations, Peasants, Natives and Non-Governmentals for an Alternative Development (*"Red Costarricense Organizaciones Rurales, Campesinas e Indígenas y No Gubernamentales para el Desarrollo Alternativo"*)
TEPROCA	Experimental Production and Commercialisation Workshop for Alternative Agriculture (*"Taller Experimental de Producción y Comercialización Agrícola Alternative R.L."*)

Summary

Agroecology, just like Organic Agriculture in Europe, belongs to the huge amount of approaches, that aim to perform agriculture in a more sustainable way. The Agroecology movement, as it is named by the famous agronomist and agroecological expert Altieri, has its origins in Latin America. The approach of Agroecology is integrating ecological measures and traditional (indigenous) knowledge into the farming system, aiming to perform agriculture in a more sustainable and also to attain food sovereignty and food security. Regarding to the sheer size of Latin America, several regions are subdivided by having their own sub-movement (Brazil, Andean Region, Central America, Cuba, Mexico). By focussing on Costa Rica as part of Central America, this case study aims to give a proper understanding how the Central American movement works. The sub-movement in Central America is called the farmer-to-farmer (CAC) movement. Here, knowledge and technology are passed on between several farmers within one region, seeing themselves as peer. Trainers, so called (Agroecological) promotors are installed, are doing the preparatory work, as they are the link between knowledge and the soon-to-be Agroecological farmers. Workshops are held, and a pre-structuring is given, to enable and simplify the application of Agroecology (Altieri and Toledo 2011). CAC can be rather seen as a key methodology, having its origins in Central America. For the Agroecological farmers, the CAC approach is not to be categorised nowhere near Agroecology. CAC is just a method, whereas the farmers are practicing Agroecology as a philosophy. Their philosophy is including culture, traditional practices and community living, together with exchange (CAC) and political advocacy. Public instruments and structures are not given solely for Agroecology, but for several sustainable agriculture approaches together. A law on Organic Agriculture (including Agroecological methods) was implemented in 2007. Organic Agriculture should not be equated to Agroecology. NGOs and public bodies, that are working on the field of sustainable agricultural approaches, are aiming to unite all different systems, e.g. Agroecology and Organic Agriculture. NGOs are aiming mostly for political power, coming with one big movement, while the government is rather interested in the economic aspects of a huge sustainable production. To the farmers, Organic Agriculture was introduced to be a way to gain price premiums, by just adding seals to their product. For small-scale farmers, the main markets for Organic products (USA & EU) are not accessible. For them, certification and conversion costs are much too expensive and price premiums cannot be attained at domestic (local) markets. Many farmers are therefore looking for more sustainable and secure alternative agricultural practices, as it is Agroecology.

1. Introduction [along with Tainná Viana]

In the past 40 years, the Agroecology movement was formed in Latin America (Altieri and Toledo 2011). As defined by Gliessman (1998), Agroecology "is the application of ecological concepts and principles to the design and management of sustainable agroecosystems". From the European point of view, Agroecology seems to be somewhat between a movement, an agricultural approach and a methodology. The sheer size of Latin America and the vast number of sub-movements, as well as the lack of specific standardisation by certification bodies make Agroecology appear to be impossible to see through (Sylvander et al. 2006).

On the other hand, there is Organic Agriculture (OA), one of the first social movements in agriculture, food and nutrition. It follows the norms of the International Federation of Organic Agricultural Movement (IFOAM) from 1972 and its purpose is to design biological processes and to develop ecological systems for agriculture. As well as Agroecology, Organic Agriculture emerged as a response for the agricultural transformations during the Green Revolution. Moreover, both of them share similar goals, such as contribution to the food security and the environment conservation (Abreu et al. 2012).

The project is aiming to contribute to a better understanding of the Latin American Agroecology movement. Therefore, two countries, Brazil and Costa Rica, were chosen in order to give different perspectives about the topic.

The movement in Brazil emerged in opposition to the agriculture modernisation and as a consequence of an excluding agrarian policy, which did not provide access to government subsidies in order to contribute for the small family farms' development (Brandenburg 2002). The movement was promoted by politically engaged and NGOs and later, by farmer's organisations and the Landless Workers' Movement (MST) which was important to disseminate the ideology and practices within the country (Altieri and Toledo 2011, Brandenburg 2002). Brazil had a significant expansion of the agroecology movement since its beginning, compared to other countries and it is a pioneer in public policies, however, it still has several obstacles to overcome in order to promote equity in rural development (Caporal and Petersen 2012).

In the following report part, the Costa Rican Agroecology movement will be analysed in its structure and aims. Costa Rica, as part of Central America, belongs to the farmer-to-farmer (CAC) movement. The Agroecology pioneers and contemporary actors will be introduced and evaluated. The relation of Costa Rica to Agroecology and to the CAC movement and the differences between the two movements will be rolled out. Furthermore, a short introduction to the Costa Rican Organic

Agricultural sector will be made. Both approaches, Agroecology and Organic Agriculture, will then be compared to one another. At last, the introductory hypothesis will be discussed and reviewed.

Hypothesis: *"Agroecology in its current shape is not only a revolutionary movement, but also a valuable approach and method to manage sustainable agroecosystems. Its reputation and application are going to increase in the future."* (Moritz Stüber, 08.01.2018)

2. Methodology

The case study "The Agroecology Movement in Latin America" is a literature review, based on two sources. One part is based on an internet research, involving databases, such as "ScienceDirect" or "ResearchGate", but also informative material of local organisations and newspaper articles. The research was performed in English and Spanish. The keywords are *"Agroecology"*, *"Organic Agriculture"*, *"Sustainable Agriculture"* and *"Costa Rica"* combined with different spelling and translated versions. The websites and publications of important associations, communities and organisations also played an important role.

The second source has been an expert interview, providing a huge knowledge about the local relationships. The chosen expert was Mr. Alexander Vargas Garro, former student of the UCR (University of Costa Rica) and employee of the CUDECA - ECOS (*"Culturas y Desarrollo en Centroamérica"*), a non-profit organisation, specialised on culture and rural development. During his work at the CUDECA – ECOS, Mr. Alexander Vargas Garro did not only receive expert knowledge about the respective country - Costa Rica -, but also about the whole Central American landscape, which allows him to not only speak about one country but also to compare and relate Costa Rica to its neighbouring states. As a consulting expert, Mr. Alexander Vargas Garro already worked with important developing actors from Mexico, Guatemala, Honduras, El Salvador, Nicaragua, Costa Rica, Panama, Peru and Cuba. Therefore, Mr- Alexander Vargas Garro is the perfect expert to talk about rural development and to evaluate the Agroecological movement and its structures.

3. General information about Costa Rica

Costa Rica is one of the highest developed countries in Latin America. Its HDI, provided by the United Nations Development Programme (2016), is much higher than the in the neighbouring countries. Together with Panama, Costa Rica is located in the southern area of the Central American bridge, that connects North and South America. At the same time, the two countries count as the wealthiest in Central America. The Costa Rican HDI estimates nearly 0.8 (Fig. 10;

Annex), while other relatable countries like Nicaragua or Guatemala, are staying below 0.65 (United Nations Development Programme 2016). Especially concerning ecological measures and sustainability, Costa Rica is setting trends. Costa Rica was able to set its energy household to nearly 100% eco-power self-sufficient throughout the past years (MINAE 2015). Other, less developed parts of infrastructure, like the waste system are aimed to be improved. Costa Rica wants to be the first country worldwide to prohibit disposable plastic (Soto 2016).

Costa Rica is also one of the pioneers in Organic Agriculture across the Central American region. Organic food export and people's ecologic awareness have been rising in the past years (Barquero 2013). One of the principal promoters of Organic Agriculture is the CEDECO (Educative Corporation of Development in Costa Rica). Organic Agriculture is taking a market niche in Costa Rica since the 1980's. Even though, Costa Rica was one of the pioneers of Organic Agriculture in Central America, Organic production is estimated to be only 1.6% of the total agricultural production. More than two third of the Organic production are meant for the export (OECD 2017). The organically managed area estimates around 8,000 ha, whereas the comparable surrounding countries, like Panama or Guatemala, have around twice the Organic area of Costa Rica. The neighbouring country Nicaragua had even accounted four times the Organic area of Costa Rica in 2015 (FAOSTAT 2016). This development is not limited to the Organic sector, as Costa Rica's agricultural sector is declining in general. Figure 1shows, that the share of agriculture in the GDP was nearly reduced to a third throughout the past 20 years in Costa Rica. In 1995, the agricultural sector accounted for 14%, whereas in 2016, agriculture was estimated to value only 5.5%. The decline is not exclusive to Costa Rica, since in neighbouring countries like Nicaragua or Guatemala, the agricultural GDP share has also been reduced throughout the past 20 years (World Bank and OECD 2016). However, the economic importance of the agricultural sector has been reduced much more in Costa Rica. This is on the one hand caused by uprising new sectors like tourism (Banco Central de Costa Rica 2016) and on the other hand the result of a functioning agricultural transformation. SAPs (Structural Adjustment Programmes) and the introduction of agricultural institutions in Costa Rica in the 1980's and 1990's were changing the sector completely (Vargas Garro 2018).

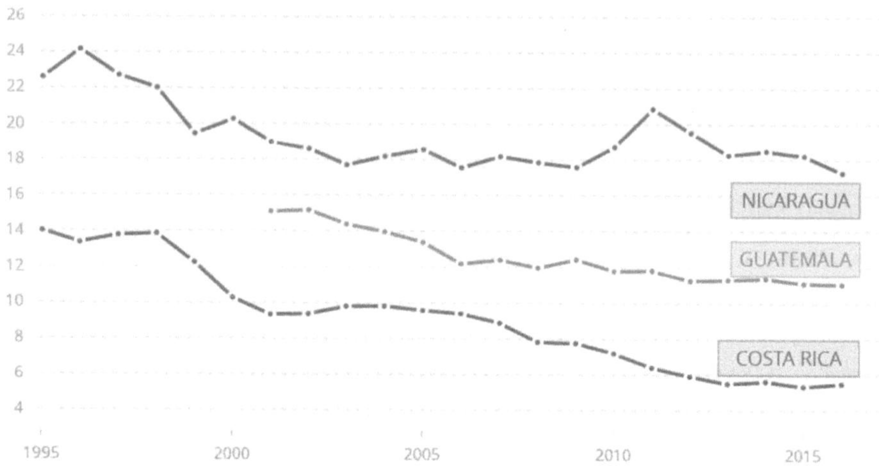

Figure 1: Share of the Agriculture sector of the country's GDP (in %) (World Bank and OECD 2016)

By means of the agricultural institutionalisation, a demographic move from rural to urban life has happened in Costa Rica, which was slowed down by civil wars in other Central American countries (UCDP 2018). The industrial level as well as the development of the service sector in Costa Rica is much higher than in the adjacent countries (Vargas Garro 2018). Even though, Costa Rica's economy was severely hit in the early 1980's (Molina Jiménez 2012), no civil war broke out like happened in Nicaragua or Guatemala (UCDP 2018). The conflicts had their origin in agricultural disputes, which were going on until the early 90's, strongly affecting rural areas. In Costa Rica the agricultural sector was restructured, and the rural area completely changed, while at the same time in Nicaragua and Guatemala, the rural area remained unchanged or even suffered setbacks (Vargas Garro 2018). The described development is therefore also reflected in the country's HDI (Tab. 1; Annex).

Concerning nature, Costa Rica has much to protect, since its nature accounts for having the highest density of biodiversity worldwide (Obando-Acuña 2007). Therefore, taking ecological measures like promoting sustainable agricultural approaches is very important to protect the country's natural treasures. Besides the in Europe commonly known Organic Agriculture, there are also other approaches in Latin America, e.g. the Agroecology movement. The Central American followers of Agroecology movement are named to be the "farmer-to-farmer" ("*Campesino-a-Campesino*" = CAC) movement, which also includes Costa Rica. Farmers in the CAC movement identify themselves as a social movement of sharing knowledge from farmer to farmer. The renowned Agroecological expert Miguel Altieri stated the CAC movement to be the *"most efficient, cheap and stable way of producing food per unit of land, input and labor"* across the region (Altieri and Toledo 2011). The movement had its origins in the late 1980's, when Guatemalan peasants visited farmers in the centre of Mexico. Each side was not persistent in their way of farming but proposed to experiment on a small-scale. The key was to stay open minded to suggestions and full of respect for the knowledge of each farmer about their own land and climate (Altieri and Toledo 2011, Holt-Gíménez 2006). The effects were astounding. Farmers in Central America were able to recover their degraded soil and at the same time increase their yield (Holt-Gíménez 2006). At some sites, the yield had even tripled. Besides the agricultural improvements, also social structures and solidarity between the farmers were rising (Altieri and Toledo 2011).

A big issue of the CAC movement is its way of communication. Since a substantial part of this movement is between the farmers themselves, research centres, universities and institutes are less integrated. Field trials were implemented recently, for example by the research centre CATIE (Tropical Agricultural Research and Higher Education Center). As a social movement, Agroecology is more popular in the northern countries, such as Nicaragua and Guatemala. Agroecology in Costa Rica is rather small-sized. It seems, that in Costa Rica, the difference between Agroecological principles and Agroecology as a movement has still not been figured out (Pou 2017). But thanks to the country's ecological awareness and many well-suited agricultural products, its potential is huge (Cortés Granados 2011; Suchini Ramírez 2012).

4. Agroecology

The Latin American Agroecology movement MAELA (*"Movimiento Agroecológico Latinoamericano y del Caribe"*) is including several Central American countries, among them Costa Rica. It is clearly declared to be a political movement, opposing Neoliberalism and Economic Globalisation. In the 1980's, first initiatives and experiences have been made in several places of Latin America, that were later formed into rural development projects, including the work of NGOs and agricultural training centres. The MAELA further states to contribute to sustainable human development by encouraging the Agroecological approach of agriculture and local knowledge (MAELA 2008a).

The formulated objective is to *"contribute to the process of social and political changes, that enable the construction of a new model of development, that is sustainable, with social justice, recovery and conservation of our ecosystem for our people (contribuir al proceso de cambios sociales y politicos que poissibiliten la construcción de un nuevo modelo de desarrollo, que sea sostenible, con justiciar social y recuperación y conservación de nuestros ecosistemas para nuestros pueblos)"* (MAELA 2008a).

MAELA can be seen as the umbrella organisation, which is coordinating and supporting many movements and communities in Latin America. In 2008, more than 150 institutions and organisations were following the movement. Many small-, mid-sized and family farmer organisations, indigenous communities and consumer associations are included. The movement supports the women's and adolescents' rights in rural areas and landless communities (FAO 2015, MAELA 2008a).

The structure of the MAELA movement is indicated in figure 2. The organigram indicates, that the MAELA is a structured organisation of international format down to its national coordinators and their respective regional farmer organisations, communities etc. For the Central American region, the respective coordinator is the *"Red COPROALDE"* (Costa Rican Network of Rural Organisations, Peasants, Natives and Non-Governmentals for an Alternative Development), with their current chairman Mr. Juan Arguedas (MAELA 2008b). The COPROALDE was founded in 1988 and is currently residing in Costa Rica. The COPROALDE network devoted itself to sustainable agriculture, food sovereignty, food safety and the CAC methodology. Furthermore, agro-ecotourism, biodiversity management and artisanal crafts are encouraged (Gloobal 2009).

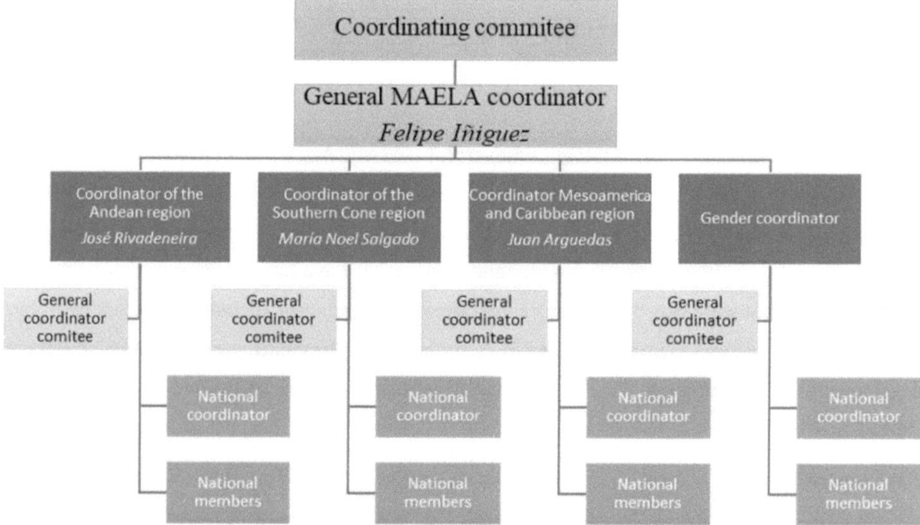

Figure 2: Organigram of MAELA (2008a, own editing with Word SmartArt)

Regarding agricultural practices, the network controls and gives access to resources, such as seeds, local fairs, soil, etc. Another important aspect is the development of a participatory certification, that enables market access for many small-sized and indigenous farmers. The COPROALDE is also taking part in creating awareness for women's rights, especially concerning rural development and agriculture (Red COPROALDE 2009), which exactly matches the conviction of its umbrella organisation MAELA.

For every region, the MAELA also receives support in terms of strategic advisory and administration. For the Central American region, Luis Samandú from the non-profit organisation CUDECA is taking the administrative position as a strategic advisor (Vargas Garro 2018). The role of Samandú can be described as a consultant. Usually, such structures and connections are staying unclear to the general public. The complex structures are not to be seen as secretive employments but are rather the cause of a lack of official structures and organisation.

Like already mentioned before, the COPROALDE network is also strictly linked to the implementation of the farmer-to-farmer methodology. Their wording is clearly indicating Agroecology as a movement and CAC as a methodology. To understand, why the COPROALDE is referring to Agroecology as a movement and to CAC as a method not a sub-movement, it is necessary to have a closer look at the CAC.

4.1. Farmer-to-farmer movement

Agroecology in Costa Rica, as part of Central America, belongs to the countries in the CAC movement (Altieri and Toledo 2011). The movement has its beginnings in the highlands of Guatemala, where farmers from the indigenous *"Kaqchikel"* Maya (Maxwell 2006), decided to visit Central-Mexican farmers in Tlaxcala in the late 1990's. The Mexican farmers founded a water and soil conservation institution. In a student-teacher-relationship, the farmers exchanged ideas, knowledge and technologies. Suggestions were implemented in small-scale test trials, respecting the traditional land and climate experiences of the hosting farmers. In the following years, especially farmers from poor, war-torn countries joined the CAC movement, spreading from the south of Mexico across all Central American countries. Only few years later, first structures were visible. In Nicaragua, the CAC method was introduced to the Nicaraguan public agricultural organisation UNAG (National Union of Farmers and Ranchers of Nicaragua) (Holt-Gíménez 2006). Even though, the movement was created in Guatemala with close relationship to Mexico, it was the UNAG, that took advantage of the CAC methodology and spread it across the country (Gonzálvez et al. 2015). Costa Rica was part of the movement since the end of 1990's, only shortly after it became spread across Central America thanks to the UNAG and other following institutions.

The Nicaraguan farmer-to-farmer programme PCaC (*"Programa Campesino a Campesino"*), founded by the UNAG, was a pioneering effort in promoting the idea of a horizontal technology transfer between farmers and Agroecological measures for soil fertility and conservation, e.g. the use of compost (Gonzálvez et al. 2015). The UNAG refined the PCaC by means of starting to train Agroecological *promotores* (Spanish: promoters) who organise and give workshops. These *promotores* are now the fundamental base of the Agroecological CAC movement, since they are the ones, responsible for passing on knowledge and technology. Until the year 2000, already 1,500 *promotores* were installed, teaching to one third of all farmers in Nicaragua (Holt-Gíménez 2006). The *promotores* themselves were regional farmers, already using available but alternative technology. Another typical property of the *promotor* principle is the absence of scientific researchers and institutes since information is taught by farmers to farmers. Agroecology-practising farmers were able to decrease soil erosion, to recover degraded land (Bunch 1990) and lower the utilisation of chemical fertilisers, while at the same time, increasing their yields. Thanks to these astounding effects, the movement spread rapidly (Holt-Gíménez 2006). By 2011, approximately 10,000 farmers and their families were practising and committing to the CAC method (Altieri and Toledo 2011). The commitment of *promotores* to support small- and mid-sized farmers can be seen as an approach, perfectly suited for rural areas, in consequence of the lack of public support and technological knowledge.

Agroecology is clearly described to be a movement, with certain structures, organisations and public programmes. Even though, farmers also have faith into the capability of CAC approach, it should rather be considered a technology. It is a method of spreading and transferring knowledge throughout a region and but not exclusively used in Agroecology. It therefore does not rely on structures but on its application to stay a commonly utilised farmer practice. In literature, CAC is described differently, either as a movement (within Agroecology) or as a methodology. A movement implies much more than just the application of one method (Vargas Garro 2018). On the contrary, Agroecology is vitally important to the farmers committing to it. Agroecology can be understood as "*a philosophy, involving culture, production, history, biodiversity, practices*" (Vargas Garro 2018), which will be examined closer in 4.2. Therefore, throughout the second part of the report, CAC will not be addressed as a movement within the movement Agroecology, but as a commonly practiced method that belongs to the Agroecology movement.

4.2. Agroecology in Costa Rica

Regarding environment aspects, Costa Rica is a country of extreme opposites. It is considered to be a pioneer of Organic Agriculture in Latin America and a role model in ecological measures. On the other hand, Costa Rica is also the world's largest conventional pineapple exporter. Conservation efforts and national parks are alternating with vast landscapes of pineapple or palm oil plantations (PROCOMER 2016). The SAPs and the regarding economic forces in the early 1990's changed the scenario of agriculture and its structure in Costa Rica completely (Vargas Garro 2018). Similar to the Green Revolution in the 60's and 70's, big plantations and mono-cultural intensive cultivation dominated the landscape. The use of chemical fertilisers and pesticides characterised the Costa Rican agriculture in the late 1990's. Costa Rican farmers were strongly affected by the industrialisation, either to set their aim to agricultural exports and to fulfil the high-yielding global agri-business dynamics, or to not be able to ensure their livelihood by farming anymore (Vargas Garro 2018). The need for counter movements like Agroecology was growing bigger during the 90's, since many farmers were opposing the development to a more intensive agriculture. Agroecology, even though it entered Costa Rica nearly at the same time as in Nicaragua, developed in a minor scale. Agroecology in Costa Rica has not received the necessary public support in its early phase, like it did in Nicaragua, since the agricultural sector at the time was expected to go into a different direction (Vargas Garro 2018, Banco Central de Costa Rica 2016). However, the minor distribution of the movement is not automatically implying a qualitative difference between the Agroecology practicing farmers in Costa Rica and in the adjacent countries. The Agroecology movement in Costa Rica can therefore still be seen as representative to whole Central America.

To understand the Agroecology movement, the complexity of its structures and the different involved stakeholders, with completely divergent perspectives, must be reviewed. From a farmer's point-of-view, who is really convinced by the practices he is applying, Agroecology is more than just an agricultural approach. The follower and the movement itself "*is eminent politic, the advocacy to generate laws, programmes and budget is without a doubt a politic action*" (Vargas Garro 2018). Across the whole country, farmers founded their own producer organisations. Together with governmental (MAG, PNAO, SIMAS) and non-governmental organisations (ANAI, ANAO, CEDECO, MAOCO), these three parties can be described as the main actors of the Agroecology movement (Vargas Garro 2018).

In the following sections (4.2.1.-4.2.3.), the three actors, their vision and aims will be outlined. It is obvious, that even within these parties, individual opinions and aims are differing widely, as well as they overlap between the three actors. However, what is reflected in the following, can be considered as the general consensus of each main actor.

4.2.1. The farmers' point of view

A farmer, who is part of the Agroecology movement, is not only applying sustainable agricultural and CAC methods. For agroecological farmers, it is a philosophy and a way of thinking, that are involving several aspects and exceeding the conventional way of practicing agriculture. The farmer's main role still is the role of a producer, but he is also carrying out culture, traditional practices and community living, with exchange and political advocacy.

It should not be assumed automatically, that all peasants are agricultural producers, farmers and vice versa. Farmers that are producing part-time in Europe are generally degraded to just pursuing a hobby. For a peasant, who is practicing Agroecology, the agriculture is not his profession, but his passion. Selling the yield or gaining profit is not the fundamental aspect of the Agroecology movement. A peasant, who is part of the movement, does not simply follow a production strategy, nor is he demanding international structures, like Agroecology seals or standards. The way of practicing is often used just to feed the own family or the surrounding, in accordance with nature.

In Guatemala for instance, indigenous people play a major role in Agroecology. Most indigenous peasants are not interested in entering a specific market or pursuing big profits, but in food sovereignty and food safety. The Costa Rican population accounts only 2.5% indigenous people, the lowest population share of all Central American countries (Schliemann 2012). Still, many indigenous Costa Ricans are practicing Agroecology. They have no intention in accessing a bigger market, moreover they are practicing *"trueque"* (Spanish: barter) (Ortíz 2016). In the case of peasants starting to trade and exchange goods between one another, their Agroecological production and turnover is not measurable, neither is the quantity of followers and producers. Of course, the exchange of goods between farmers is not limited to the indigenous population but is a common practice in the Agroecology movement. Apart from goods, Costa Rican peasants often share and exchange their knowledge. This can be considered the CAC method. However, not all farmers, who are practicing CAC methods are automatically practicing Agroecology or convinced by the Agroecology movement (Vargas Garro 2018). The farmers are not only exchanging knowledge, but come together to form communities, similar to producer organisations. Uncountable producer organisations have already been founded in Costa Rica and across Central America, with no overview possible, since only a few of them does also exist in bureaucratic terms.

Since the Agroecology producers deeply believe in their way of farming, they also act together as a political community. Together as a community, the farmers openly confront big agricultural and pharmaceutical corporations like Monsanto, large-scale plantations and mono-cultivation. Within local communities, the term "Agroecology" is often replaced with "Sustainable Agriculture", "Climate Smart Agriculture" or other names, which also contribute to their identity. Often, the different terms do not change the fundamental belief of uniting agriculture and ecology in the most sustainable way.

The insufficient data, regarding producer associations and the turnover, as well as the sub-division into different names, all brought together in one movement seems to be elusive. Nevertheless, this elusiveness is also reflected in the aim of the Agroecological peasants. They are part of an opposition movement, with no interests in profit and conventional structures, but political influence and an improved sustainability management.

4.2.2. The governmental organisations' point of view

In recent years, the government has started to incorporate the Agroecological approach into their public policy. The government had been reacting to the Agroecology movement only after several success stories took place in Costa Rica and in the surrounding countries (Altieri and Toledo 2011). Only 2007, the Ministry of Agriculture and Livestock (MAG) released official directives, which included Agroecology. The MAG's "Department for Promotion of Organic Farming Production" (DFPAO) formed the Law No. 8591 on Organic Agriculture, subrogated by the "Department of Sustainable Production" (further information in 5.2). In the law, the MAG is also mentioning technologies, designed by the Agroecology movement. The RBA (Recognition of Environmental Benefits) and the RBAO (Recognition of Environmental Benefits for Organic Production) can also be considered as public incentives on practicing Agroecology. On the contrary to the Law on Organic Agriculture, the RBA and RBAO are no regulation neither a legislation, but as incentives to generate knowledge about the system of production and the financial aspects (Sáenz-Segura et al. 2017, IBS Soluciones Verdes 2013).

The SIMAS (Mesoamerican Information Service on Sustainable Agriculture) promoted Agroecology, by means of their Agroecological and Organic Producer movement (*"Movimiento de Productoras y Productores Agroecológico y Orgánico"*) and their Agroecological Alliance (*"Alianza por la Agroecología"*), a platform for sustainable rural development. SIMAS is a civil association, legalised by the Law No. 8408 on Sustainable Production by the MINAE (Ministry of Environment, Energy and Telecommunications) in 2003 (SIMAS 2018, Sáenz-Segura et al. 2017). Furthermore, the Nicaraguan governmental organisation UNAG backed up Agroecology with scientific researches, also related to Costa Rica (Vargas Garro 2018).

It is important to point out that the government's interests clearly lie on the Agroecological approach, not on the Agroecology movement. All actions undertaken by the government are aiming to pick up the methods and convert them into a sustainable and profitable business, for instance amalgamated with Organic Agriculture. Even though, political advocacy is of interest to the peasants, the way of certification and export orientated production, as it is aimed by the MAG (OECD 2017), is opposing their philosophy.

4.2.3. The non-governmental organisations' point of view

NGOs are mostly focussing on research and the scientific fundament of Agroecology, such as the CEDECO. If a method proves itself to be use- or successful, it is then multiplied in the surrounding area and spread throughout the institutional system (Vargas Garro 2018). For instance, researches by MAOCO (Organic Agricultural Movement) and ANAO (National Association of Organic Agriculture) have been contributing to the Law on Organic Agriculture, in terms of scientific evidences, years of incidence and the compensation of public absence (2000-2006), before the law finally had been approved in 2007 (Sáenz-Segura et al. 2017, IBS Soluciones Verdes 2013).

The ANAI association is playing a similar role. Already founded in 1983, the ANAI originally was dedicated to nature conservation in the Talamanca region in the southern region of Costa Rica. It later broadened its aim to also *"helping people put into practice community and landscape-level initiatives that integrate nature conservation and the well-being of the people who conserve and sustainably use nature"* (ANAI 2017a). Their aim also matches with the application of Agroecology which is performed in ANAI training programmes. Within their Agroecology Programme, the ANAI claims to have already trained thousands of eco-volunteers and interns in on-the-ground research and conservation work. This system is similar to the CAC method, since many of the trainees eventually spread their gained knowledge in future conservation and development programmes all across the world (ANAI 2017b). The ANAI is the partner organisation of the APPTA community (further information in 4.2.4.). The APPTA community is a small producer association, founded in 1987, that is practicing Organic Agriculture, together with Agroecological and sustainable methods (APPTA 2017).

Organisations like MAOCO, ANAO or corporations like the ANAI were advocating all different kinds of ecological approaches in their early stages of development in Costa Rica. Not only did they support the development of Agroecology, but also of Organic Agriculture. They count as the major organisations, which promote and do research on Agroecology. Apart from the opposing philosophies and interests of the different approaches, their goal is to create a combined (political) force to impose the way of sustainable agriculture (Sáenz-Segura et al. 2017).

Corporations that are working with big plantations generally oppose the Agroecological approach. *"Naturally, these approaches* (Agroecology) *have many enemies such as big plantations companies like Monsanto, ADM, Carrefour etc., who based their production in huge pieces of land, even against the communities, laws etc"* (Vargas Garro 2018). These big companies do political advocacy in Costa Rica in their interests, just as the Agroecological farmers do. Many small-scale farmers have realised the unsustainable approach of agriculture, which these companies are promoting. Many farmers feel the urge to not only stop participating but to politically oppose this unsustainable way. *"In this sense, some kind of articulation will be always needed to oppose such invasive projects and to spread more sustainable ways to produce food"* (Vargas Garro 2018).

4.2.4. Examples

The CEDECO is taking a major role in the Costa Rican Agroecology movement, by using their own approach *"Cadena Agro-Alimentaria"* (Spanish: Agri-food Chain), which integrates small farms and jointly forms an Agroecology perspective, including most components of the production chain. Starting from the elaboration phase of an Agroecological farm, passing through the processing stage of agro-industrialisation until reaching the market, the CEDECO wants to offer a business approach, that is available for everybody. Peasants and farmer communities are enforced to enter the bigger market, consulted in the elaboration of marketing strategies and provided with information about professional quality management. Their project is offered online to all peasants in Central America, that have already more than two years of Agroecological experience (CEDECO 2012b, Vargas Garro 2018). The CEDECO is also supporting the company CoopeAgri El General R.L. The company, which is mainly distributing Organic coffee and sugar, was already founded in 1962 to solve the problems, that came with the commercialisation and industrialisation of the coffee production. It currently has more than 10,000 associated producers and is a pioneer company, which has promoted economic growth in the extensive southern area of Costa Rica as well as sustainable production (CoopeAgri 2008). The certified Organic production is also including Agroecological methods. Their sugar production was certified Fairtrade in 1998 and coffee in 2004 (Fairtrade 2018). CoopeAgri enables small-scale farmers to enter a bigger market by simultaneously aiming to benefit the producers in social, environmental and cultural terms.

The last indicated example is the APPTA association, which was founded in 1987. APPTA is working with 932 small-scale producers in 56 different municipalities in Talamanca in the South-Caribbean region in Costa Rica (APPTA 2017). 80% of the members are indigenous producers, therefore APPTA is contributing to bringing back traditional indigenous methods of agriculture, like polycultures and agroforestry. Through efficient production, associates training and commercialisation, APPTA aims to undertake steps to improve the associate's quality of life, protect the environment and conserve the natural, ethnic and cultural wealth of Talamanca and surrounding regions.

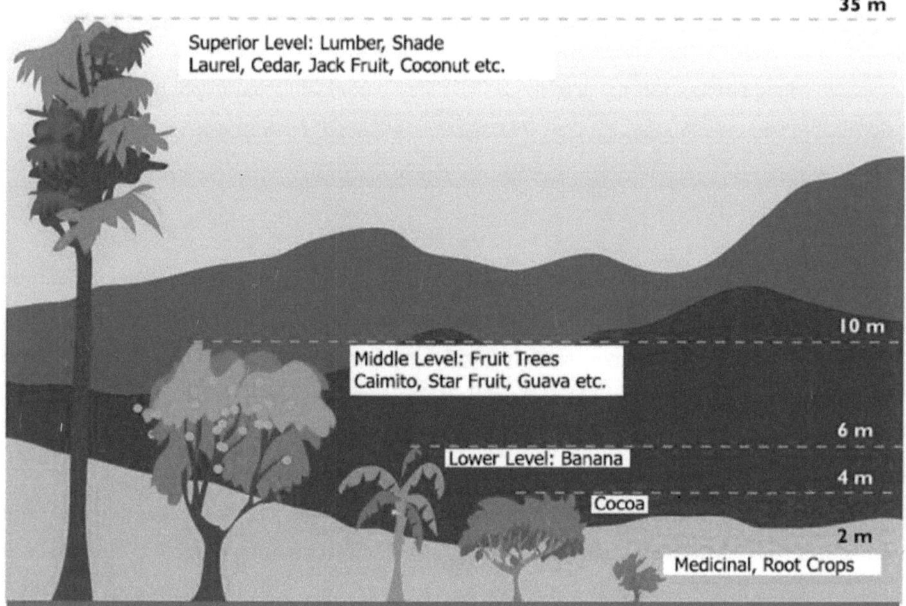

Figure 3: Agroforestry system designed by the APPTA association, perfectly suited for their associative peasants (APPTA 2017)

The range of products differs between dried cacao to several fruits (banana, guava, star fruit) and is provided in one system of agroforestry (Fig. 3). All associative peasants are contributing together to the development of each agricultural system (APPTA 2017). Two descriptive videos as well as further pictures of the agroforestry systems are indicated in the Annex (Tab. 2).

The association is giving workshops on Organic Agriculture methodology and other environmental approaches, like Agroecology (Fig. 4). Treatments, that require external knowledge or technology is provided to the peasants by APPTA as well. For instance, grafting the cacao trees or the production of Organic compost are either trained or implemented by the association (APPTA 2017). APPTA is also abiding the strict guidelines of the Eco-LOGICA association to assure product quality and access to the Organic market (Eco-LOGICA 2012).

The huge amount of certification is helping the Agroecological farmers to access a global market (Switzerland, EU and USA) as it is indicated in figure 5, without carrying the costs of certification on their own.

For their products, APPTA has its own logo (Fig. 4).

Figure 4: Logo and seal of the APPTA association (APPTA 2014) and a workshop, held on Organic Agriculture methodology (APPTA 2017)

Besides their own logo, the products of APPTA are also labelled by a range of certification bodies (Fig. 5):

Figure 5: All seals, the products of the APPTA association are certified with (from left to right: Organic Certification of USA; Organic Certification of Costa Rica; Fair Trade Certification of Europe; Organic Certification of Switzerland; Fair Trade Certification of USA)

The APPTA association and their associated peasants are definitely contributing to the Agroecology movement, even though, they are closely linked to Organic Agriculture as well. Their advantage is the combined production as part of the association and therefore lower costs and difficulties. To understand the peasants' situation, who are stuck between a huge amount of certification bodies, a closer look on Organic Agriculture has to be made.

5. Organic Agriculture in Costa Rica

5.1. History

Costa Rica's early Organic production trials date back to 1980, as one of the first Central American countries. In this year, several small-scale farmers associated their horticultural production and commercialised to the organisation TEPROCA (Experimental Production and Commercialisation Workshop for Alternative Agriculture). In 1984, they officially founded this first Organic organisation in Costa Rica in cooperation with the Japanese organisation JICA, that supported the capacity and the first Organic trials (IBS Soluciones Verdes 2013, CEDECO 2001). The CEDECO itself, as one of the most important organisations of rural development in Costa Rica, was founded in the very same year (CEDECO 2012a). Only three years later, the national aim of an export orientated Organic production was formulated. Banana and cacao were the first commercialised Organic products to enter the European market (García 1998). The crucial development of Organic Agriculture in Costa Rica started in the 1990's, therefore contemporaneous to the EU's Organic development. In 1990, public policy instruments, laws and national programmes regarding Organic Agriculture were implemented. Since 1994, Costa Rica also has its own National Programme of Organic Agriculture, by means of an executive decree of the MAG. The last implementation step was made in 1998, when the official Costa Rican organically certified label was introduced by the Agricultural Ministry (Fig. 6).

Figure 6: Latest version of the Costa Rican Organic label (left) and the seal (middle) by the Ministry of Agriculture and Livestock (right) (MAG 2014; APPTA 2017)

Because of the government's implementations, several farmer associations were founded in the 1990's, for instance the Costa Rican Organic Agricultural Movement (MAOCO), that was founded in 1999 (IBS Soluciones Verdes 2013).

5.2. Organic Agriculture in Costa Rica

Throughout the following years, the Organic sector did not change that much. The Organic area and the market share of Organic production fluctuated over the years. In 2015, the Organic production area was estimated 11,000 ha (Fig. 8) (PEN 2016) or 8,000 ha (Fig. 9) (FAOSTAT 2016). The difference might come from different statistical methods, whether all Organic or only IFOAM-certified areas are taken into account.

Considering the 8,000 ha of organically managed land in 2014, this accounts for only 1.6% of Costa Rica's total agricultural land (~450,000 ha). The Organic share of the total agricultural production is therefore below world average (OECD 2017). In the Costa Rican conventional agriculture, pineapple is the principal arable crop. With a global market share of 55%, Costa Rica is even the leading exporter of pineapple, worldwide (PROCOMER 2016). Pineapple as the principal crop (24%) is followed by banana (18%), coffee (6%) and sugar (5%) in annual production (Fig. 7). Pineapple counts as the most intensively farmed crop in Costa Rica. Therefore, less area is needed for a higher production rate. Concerning the agricultural area, the rather extensive coffee production is leading with 22%. The whole area of pineapple cultivation accounts for only half of the coffee cultivation area (MAG 2010). In Organic Agriculture, banana cultivation is the leading crop (37%), followed by cacao (21%) and pineapple (18%) production (Fig. 7) (MAG 2016b).

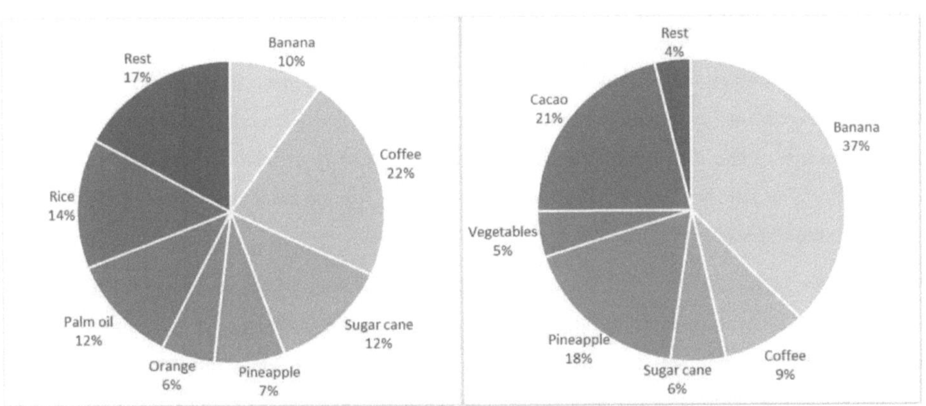

Figure 7: Total (left) and Organic (right) area of agricultural production in Costa Rica (in %) (MAG 2016b, MAG 2010)

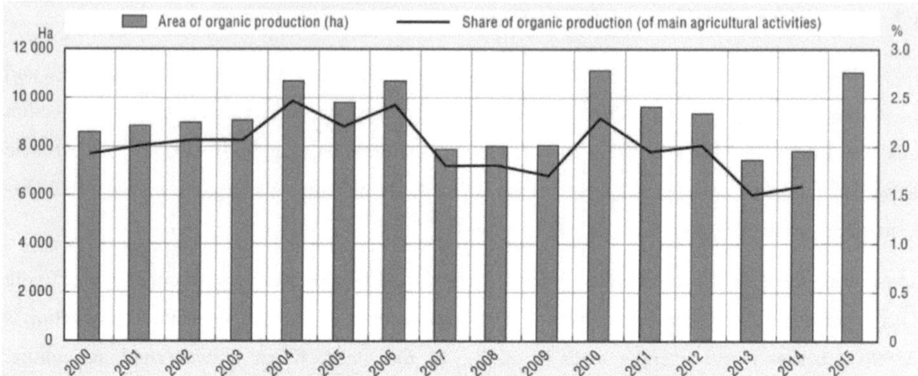

Figure 8: Area of Organic production and share of Organic production of Costa Rica's total agricultural production (OECD 2017, PEN 2016)

It is possible to identify two drops in both, the Organic area as well as in the Organic market share (Fig. 8). The first drop took place between 2007 and 2009, as Costa Rica was severely hit by the World Economic Crisis. Fluctuating prices and diminishing demand from the EU and the US led to economic losses of the export-oriented country Costa Rica (OECD 2017). The second drop happened in 2011 and found its low point in 2013. The MAG, together with IBS Soluciones Verdes (2013), state different reasons for the fall of Organic production. Organic farmers in Costa Rica very often lack of resources and miss public and private support. They claim, that channels for marketing, distribution and commercialisation are still underdeveloped. Although, the Costa Rican Organic market is mainly export orientated, nearly 40% of Organic farmers state to have issues with finding marketing and distribution channels. A national strategy to facilitate the Organic farmers market access is still not given. This especially concerns the connection of small-scale Organic farmers to the international market.

The report by MAG (2013) matches with the results from the State of Nation Program (PEN 2015). The State of Nation Program was mainly concerned about the private and public support. Farmers are facing problems with service extensions and innovation systems. Many farmers also do not feel the confidence of receiving price premiums for Organic products. Barquero (2010) is stating Organic Agriculture to experience a slow progress in Costa Rica, affected by cultural obstacles and prejudices, even though it is proven to benefit the farmers as well as the environment.

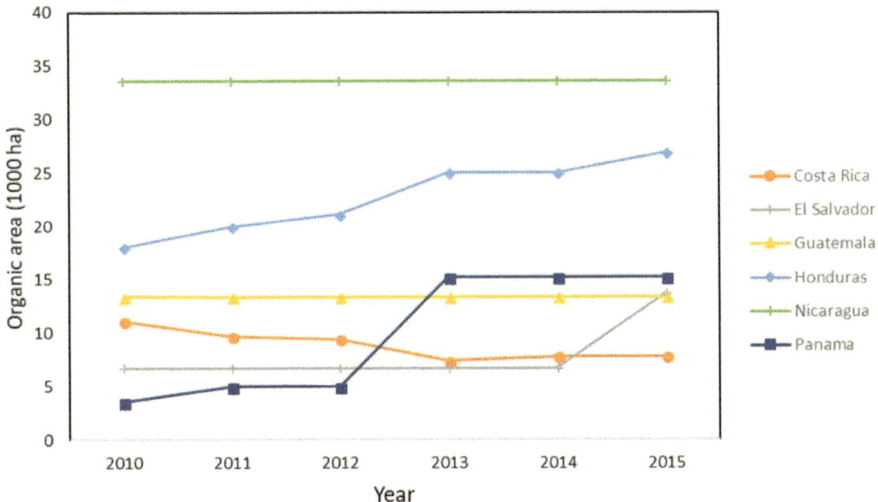

Figure 9: Organic land of Costa Rica and other CELAC countries in the Central American region (FAOSTAT 2016)

In 2016, the MAG (2016) was specifying their review on Organic Agriculture in Costa Rica. The high costs of certification and stringent requirements on traceability were criticised. Farmers are generally overstrained by the amount of data, associated with the low public support. The mechanisation and production techniques are still limited. Also, the expenses on labour and Organic fertilisers often exceed the farmers' capabilities. All mentioned reasons lead to the fact, that Costa Rica's Organic sector is underdeveloped. The Organic sector of the neighbouring countries is way bigger. Many neighbouring countries have a higher land surface and agricultural area, which of course is also playing an important role. The Costa Rican Organic sector seemed to be promoted poorly in its early stages in a time, while SAPs and its associated economic forces industrialised the agricultural sector (Vargas Garro 2018).

On the other hand, the Agricultural Public Sector, under responsibility of the MAG, is aware of the importance of the Organic market, as it is growing globally. Besides the export orientation of the Organic sector, the domestic demand is also increasing. The ecological awareness of the urban middle class, as well as their demand for high-quality, standardised food is rising (OECD 2017). Tourism counts as one of the highest growing economic sectors of Costa Rica. Lately, the revenue increased constantly and will soon exceed 4 Mio. $ annually (Barquero 2017, Banco Central de Costa Rica 2016). Ecotourism is Costa Rica's plays an important role in protecting the country's nature and biodiversity. In fact, Costa Rica counts as a pioneer and role model for ecotourism sector

(Honey 1999), too. As the sector of ecotourism grows, the hotels and restaurants also rely on a growing market for sustainable products, such as Organic. In fact, the domestic and international demand for Costa Rica's Organic products does exist. Eco-labels increase the potential of the agricultural production, thanks to the country's reputation for environmental protection, such as nature conservation, ecotourism and power management. The logical consequence are public investments and the commercialisation of the Organic market. A first step forward was undertaken in 2007, when the Law No. 8591 on Organic Agriculture was approved. The law mainly concerns about development, promotion and support of Organic farming activity, which can be described as the beginning of the institutionalisation of the Organic sector in Costa Rica (IBS Soluciones Verdes 2013). For the farmers themselves, this means an increase of public support, by means of tax provisions and financial incentives. The law mainly aims at small- and medium-scale farmers. Even tax exemptions are possible, if the company is focused mainly on the export market (OECD 2017). In 2015, Costa Rica introduced further policy instruments, that partly benefit Organic farming. Organic farming has been recognised as environmentally beneficial, which relates it to the MINAE. Thanks to the RBAO, Organic farmers can receive direct payments for their environmental service, for a maximum amount of three years.

6. Discussion

After the two individual parts about Agroecology (4) and Organic Agriculture (5), both approaches will be compared to one another in the following. Since the point of view towards other approaches is differing between the different main actors, the comparison will also be subdivided. In table 1, the three different main actors (peasants, public sector, NGOs) are presented. The public sector is represented in form of the Agricultural Ministry and their sub-department on Sustainable Production, that has developed the Law on Organic Agriculture in 2007. The law is also including Agroecological measures. The NGOs are represented by the MAOCO and the COPROALDE network. The MAOCO also contributed to the Law on Organic Agriculture in 2007 and was replacing the government the years before. The COPROALDE is currently coordinating the Agroecology Movement (MAELA) in Central America. Of course, the position and personal interests of every individual cannot be considered. What is described in table 1 is the general consensus of each respective actor.

The Agroecological vision of the peasants is far different from the other two main actors since many of them are practicing Agroecology without economic interests. Certification bodies, seals and the costs of the conversion are mostly opposed by the peasants that are practicing Agroecology as a movement and philosophy. The government and many NGOs are aiming to unite Agroecology and Organic Agriculture, but for different reasons. Organisations like the MAOCO have been working hard on promoting all kinds of sustainable agriculture approaches to enforce the political and social power. A split into different approaches or movements would also mean the split of political influence. Divided and less political influence is therefore opposed by these organisations. Many projects, implemented or supported e.g. by the CEDECO or the ANAI, are not only practicing Organic Agriculture but often also promoting Agroecology.

Table 1: Different conceptions between the three different types of main actors of the Agroecology movement in Costa Rica (peasants, MAG as a public body and MAOCO and COPROALDE, as NGOs) (Sáenz-Segura et al. 2017)

	Peasants	**MAG & Department of Sustainable Production (Government)**	**MAOCO & COPROALDE**
Agroecological vision	Agroecology is a philosophy, involving culture, tradition, (production) practices, biodiversity and sustainability. Hostile towards OA and certification generally, too expensive, not sustainable, no price premiums.	Same principles between Agroecology and OA. The only difference is the Organic certification. No interests in the *"political movement"*.	Agroecology = OA. Strategy of uniting the forces into one social movement.
Technical approach	Knowledge exchange (CAC) Political advocacy Sustainability	Promotion of sustainable agricultural production	Natural regulation of fertility. Ban of GMOs and agrochemicals.
Institutionalisation	Non-bureaucratic communities and associations Organisation internal and mostly unofficial	Sustainable (RBA) and Organic (RBAO) incentives	Political presence on a national scale, in terms of the OA Law (No. 8591). Communities and political presence at the level of municipalities One sector (political strength by uniting OA and Agroecology)

The government on the other hand is not interested in the movement's political part. While the NGOs are interested in joining political power, the government is interested in the economic force of one total movement.

A formal institutionalisation was achieved to promote Organic Agriculture, including Agroecology methods (Law on Sustainable Production and Law on Organic Agriculture). Agroecology has no proper law in Costa Rica. The Costa Rican public sector did not promote neither one of the two approaches in their early stages, which eventually lead to a smaller development. However, Organic Agriculture has a better public structure in Costa Rica than Agroecology. The Agricultural Ministry and its Sustainability and Organic Agriculture Departments are paying much less attention to Agroecology. Only the faculties of agronomy of the two most important universities in Costa Rica, UNA (National University of Costa Rica) and UCR seem to promote both approaches equally by means of scientific investigations. The UNA even has a master programme linked to R&D, which is called "Alternative Agriculture with focus on Ecological Agriculture" ("*Agricultura Alternativa con mención en Agricultura Ecológica*") (UNA 2018, Vargas Garro 2018).

The Costa Rican Agroecology movement is still weak in terms of political influence since many projects and farmer associations (such as APPTA) are still isolated. However, the low political influence may not be confused with low political interests from the farmers' side. Farmers who are practicing Agroecology are formulating law- and programme-like structures together with NGOs and local corporations, like the ANAI or the MAOCO. Main elements of their work are Agroecology trainings and the spread of such technology, which are not necessarily to be understood as a movement. What is happening in fact is, that many farmers are joining such programmes, as they see how the Agroecology approach is including economic interests but also their own welfare and nature conservation. It can be stated, that new programmes are implemented, and many farmers are implementing Agroecology, because they believe in its system (Vargas Garro 2018).

The approach of Organic sector in Costa Rica is showing a complete opposite trend. For small-scale farmers in Costa Rica, Organic Agriculture is not sustainable. While Agroecology often means less expenses, through community support, absence of pesticides and more handicraft, Organic Agriculture means higher expenses for the farmers. Price premiums through Organic conversion and certification have been promised to the them, but these prices cannot be achieved in the domestic (local) market, yet. Even though, the Organic market share was increasing in Costa Rica recently, only bigger producer can attain price premiums, which they receive in the bigger cities of the country and abroad. A long-lasting conversion process, certification, Organic pesticides, etc. is not suitable for small-scale farmers. Most farmers converted to Organic Agriculture in order to export to the big markets like in the US or EU, while regional markets are losing relevance in terms of the Organic seal, but not in terms of sustainable practices. The regional market often makes no differences between conventional and Organic products. In this sense, many disappointed former Organic Agriculture producers have moved on to more sustainable agriculture proposals. NGOs and research centre programmes are aiming to support Organic production, by giving structures and covering the biggest costs, but convincing these farmers of both approaches is a time-consuming process.

Conventional agriculture is cheap, but not environment-friendly, while Organic Agriculture is more expensive and not sustainable for the small-scale farmers. The mentioned more sustainable agriculture proposals are regionally called differently. What they show altogether, is the urgent need for alternatives: *"Agroecology and Sustainable Agriculture, or how it is now called Climate Smart Agriculture are capturing more producers and farmers. I doubt a movement "must" exist to promote this"* (Vargas Garro 2018).

For farmers in Costa Rica, Organic Agriculture is as possibility to receive a higher price and to access a new market. In an export-oriented agricultural sector, only big-scale farmers can afford the difficulties and expenses, that are involved in practicing Organic Agriculture. Farmers, who are convinced to do sustainable agriculture, for nature and themselves, are interested in less economic approaches, such as Agroecology. Agroecology farmers strongly believe in the philosophy of their movement, which they want to practice life-long and then pass on to the next generation. Farmers that are practicing Organic Agriculture are practicing it for economic reasons and will quickly move on, if the practice is not worth the effort anymore. Organic Agriculture provides more legal requirements (IFOAM regulations and national laws), but this does not automatically imply a better structured development. Exactly the huge number of legal requirements are making it hard for many farmers to practice Organic Agriculture in Costa Rica (Vargas Garro 2018). Agroecology is the better suited approach for small-scale farmers in the rural areas of Costa Rica.

7. Conclusion & outlook

Costa Rica is a reference in terms of biodiversity, conservation, low carbon economy and had been even a pioneer in Organic Agriculture but is no role model for Organic Agriculture any more. Laws and public policy programmers were first set in the late 1990's and early 2000's, which was too late to support Organic Agriculture in its early phase. While big-scale farmers benefit economically by practicing Organic Agriculture, thanks to an export-oriented policy and price premiums, small farmers have more disadvantages on their side. Years of conversion, the costs for certifications and Organic pesticides, etc. are unsustainable for small-scale farmers. Many smaller farmers are aiming to practice a more sustainable way of agriculture, which means to be environmentally friendly and to secure themselves against food scarcity and poverty. Agroecology is such a valuable approach, since many farmers are coming together in farmer communities or associations, helping each other and exchanging knowledge (CAC). This also includes the work of NGOs and research centres. They are interested in the positive effects, Agroecology has on the farmers' prosperity, nature (soil, biodiversity) and even the production rate.

In the introduction, the preliminary hypothesis was mentioned. The hypothesis is reflecting the author's expectation at the beginning of the case study.

Hypothesis: *"Agroecology in its current shape is not only a revolutionary movement, but also a valuable approach and method to manage sustainable agroecosystems. Its reputation and application are going to increase in the future."* (Moritz Stüber, 08.01.2018)

The author's hypothesis can be considered as correct. Agroecological methods have left an impression on the public sector, which is why they were then integrated into laws (Law on Sustainable Production and Law on Organic Agriculture) and projects. However, the author misjudged the relation between Agroecology as a movement and as a technology. The aims of soil conservation and nature integration can be achieved by all different kinds of sustainable agriculture approaches. What is impressive, are the strong bonds that have developed between the farmers themselves and between them and nature. Practicing a life-long philosophy of sustainable agriculture, together with food and poverty security, is much more relevant to Agroecological farmers than accessing new markets and maximising profit and production. Many farmers are expected to commit to the Agroecology movement in the future. NGOs and research centres are aiming to motivate more and more peasants to participate in the movement, especially in rural areas. Agroecology in Costa Rica and across Central America should not be seen as a new revolutionary movement, but as a step back to tradition, culture and social security, joined by political advocacy towards their vision of a sustainable, long-established agriculture.

8. References

Abreu, L. S. de, Bellon S., Brandenburg A., Ollivier G., Lamine C., Darolt M. R., Aventurier P. (2012): Relações entre agricultura orgânica e agroecologia: desafios atuais em torno dos princípios da agroecologia., Embrapa Meio Ambiente, p.26.

Altieri, M. A., Toledo V. Manuel (2011): The agroecological revolution in Latin America, Journal of Peasant Studies, 3, p.587–612.

ANAI (2017a): 40 years, https://www.anaicostarica.org/, accessed 26.06.2018.

ANAI (2017b): Overview, https://www.anaicostarica.org/overview.html, accessed 26.06.2018.

APPTA (16.12.2014): Logo de APPTA, http://www.appta.org/?attachment_id=7, accessed 27.06.2018.

APPTA (2017): Website, http://www.appta.org/?page_id=35, accessed 26.06.2018.

Banco Central de Costa Rica (19.10.2016): Turismo es el sector que más crece en la economía costarricense, https://presidencia.go.cr/comunicados/2016/10/turismo-es-el-sector-que-mas-crece-en-la-economia-costarricense/, accessed 14.06.2018.

Barquero, M. (13.09.2010): Múltiples obstáculos retrasan avance de agricultura orgánica, https://www.nacion.com/economia/multiples-obstaculos-retrasan-avance-de-agricultura-organica/QDLY52OTCREGHPYFYGEXZZAGM4/story/, accessed 12.06.2018.

Barquero, M. (10.07.2013): Nuevos negocios se especializan en vender productos orgánicos en Costa Rica, http://www.nacion.com/economia/Nuevos-especializan-organicos-Costa-Rica_0_1352864762.html, accessed 29.10.2017.

Barquero, M. (2017): Ingresos por turismo mantienen un crecimiento sostenido, https://www.nacion.com/economia/indicadores/ingresos-por-turismo-mantienen-un-crecimiento-sostenido/C4DI3BXDAJAJLNPDYZNXSHX6NQ/story/, accessed 14.06.2018.

Brandenburg, A. (2002): Movimento agroecológico, Desenvolvimento e Meio Ambiente.

Bunch, R. (1990): Low Input Soil Restoration in Honduras, International Institute for Environment and Development (IIED), London.

Caporal, F. Roberto, Petersen P. (2012): Agroecologia e políticas públicas na América Latina: O caso do Brasil, Agroecología 6, p.63–74.

CEDECO (2001): Estudio de Factibilidad: Centro de Acopio de Hortalizas Orgánicas.

CEDECO (2012a): 27 años de crecimiento y transformación, http://www.cedeco.or.cr/es/contents/view/logros, accessed 11.06.2018.

CEDECO (2012b): Cadena Agro-Alimentaria, http://www.cedeco.or.cr/contents/view/cadena-agro-alimentaria, accessed 27.06.2018.

CoopeAgri (2008): Quiénes somos, https://ddd.uab.cat/pub/notfrefas/notfrefas_a2012m1d17/ig_quienes_somos.htm, accessed 27.06.2018.

Cortés Granados, V. (2011): Agroecología del agroecosistema café (Coffea arabiga) y su relación con la erodabilidad de laderas en el Valle de Orosi, Cartago, Costa Rica., Anuario de Estudios Centroamericanos, Vol. 37, pp. 271-305.

Eco-LOGICA (2012): Certificadora de Productos Orgánicos y Sostenibles Eco-LOGICA, https://www.eco-logica.com/, accessed 28.06.2018.

Fairtrade (2018): COOPEAGRI, Costa Rica, https://www.fairtrade.org.uk/Farmers-and-Workers/Coffee/COOPEAGRI, accessed 27.06.2018.

FAO (2015): Movimiento Agroecológico de América Latina y el Caribe (MAELA), http://www.fao.org/family-farming/detail/es/c/326708/, accessed 24.06.2018.

FAOSTAT (2016): Land Use, FAO.

García, J. E. (1998): La agricultura orgánica en Costa Rica, EUNED, San José, Costa Rica.

Gloobal (2009): Red COPROALDE, http://www.gloobal.net/iepala/gloobal/fichas/ficha.php?id=28745&entidad=Agentes&html=1, accessed 24.06.2018.

Gonzálvez, V., Salmerón-Miranda F., Zamora E. (2015): La Agroecología en Nicaragua: La praxis por delante de la teoría, Murcia, Spain.

Holt-Gíménez, E. (2006): Campesino a campesino Voices from Latin America's Farmer to Farmer Movement for Sustainable Agriculture, Food First Books, Oakland, Calif.

Honey, M. (1999): Ecotourism and sustainable development Who owns paradise?, Island Press, Washington, DC.

IBS Soluciones Verdes (2013): Estudio sobre el entorno nacional de la agricultura orgánica en Costa Rica, Ministerio de Agricultura y Ganadería (MAG), p.40.

MAELA (27.10.2008a): ¿Que es MAELA?, accessed 24.06.2018.

MAELA (27.10.2008b): Regiones Y Miembros, accessed 24.06.2018.

MAG (2010): Costa Rica - Informe Final.

MAG (2014): Nuevo sello orgánico se convertirá en herramienta de control y mercadeo para posicionar los productos nacionales en los mercados más importantes del mundo.

MAG (2016a): General information on Extension Services, Costa Rica.

MAG (2016b): Servicio Fitosanitario del Estado, p.2-3.

Maxwell, J. M. (2006): Kaqchikel Chronicles The Definitive Edition, University of Texas Press, Austin, TX.

MINAE (2015): Plan Nacional de Energía 2015-2030, Ministerio de Ambiente y Energia, San José, Costa Rica, p.1-2.

Molina Jiménez, I. (2012): Crisis económica y escenarios futuros. El caso de Costa Rica (1981-1982), Boletín AFEHC, Asociación para el Fomento de los Estudios Históricos en Centroamérica, Universidad de Costa Rica, No. 53.

Obando-Acuña, V. (2007): Biodiversidad de Costa Rica en cifras, Instituto Nacional de Biodiversidad (INBio), Costa Rica, p.8-17.

OECD (2017): Agricultural Policies in Costa Rica, OECD Publishing, Paris.

Ortíz, A. (01.06.2016): Sexto Trueque por la Vida y la Sostenibilidad, http://alianzaagroecologia.redelivre.org.br/2016/07/sexto-trueque-por-la-vida-y-la-sostenibilidad/, accessed 26.06.2018.

PEN (2015): Estado de la Nación en desarrollo humano sostenible Un análisis amplio y objetivo sobre la Costa Rica que tenemos a partir de los indicadores más actuales (2015), Programa Estado de la Nación, Pavas, Costa Rica.

PEN (2016): Figure 1.14. Organic production area in Costa Rica Environmental Database.

Pou, A. (01.01.2017): La agroecología en Costa Rica: un proyecto de desarrollo cultural, http://www.revistapueblos.org/blog/2017/01/05/la-agroecologia-en-costa-rica-un-proyecto-de-desarrollo-cultural/, accessed 14.06.2018.

PROCOMER (2016): Guía Informative sobre Temas de Comercio Exterior, Centro de Asesoría para el Comercio Exterior (CACEX); Dirección de Inteligencia Comercial, San José, Costa Rica.

Red COPROALDE (2009): Mujeres rurales y propuestas de desarrollo.

Sáenz-Segura, F., Le Coq J.-F., Bonin M. (2017): Políticas de apoyo a la agroecología en Costa Rica, CINPE-UNA; Cirad-Ciat; Cirad.

Schliemann, C. (2012): La autonomía de los pueblos indígenas de Costa Rica una contrastación del estándar internacional con la legislación nacional y su implementación, Revista Latinoamericana de Derechos Humanos, 23 145. Volumen 23 (1).

SIMAS (2018): Quiénes somos, http://simas.org.ni/quienes-somos/, accessed 26.06.2018.

Soto, M. (16.11.2016): Costa Rica se encamina a eliminar los plásticos de un solo uso, www.nacion.com/vivir/ambiente/Costa-Rica-encamina-eliminar-plasticos_0_1597840214.html, accessed 29.10.2017.

Suchini Ramírez, J. G. (2012): Innovaciones agroecológicas para una producción agropecuaria sostenible en la región del Trifinio, Real Embajada de Noruega, Cartago, Costa Rica.

Sylvander, B., Bellon S., Benoit M. (2006): Facing the organic reality: the diversity of development models and their consequences on research policies, JOC Odense, Denmark.

UCDP (2018): Uppsala Conflict Data Program, http://ucdp.uu.se/#country/93, accessed 25.06.2018.

UNA (2018): Agricultura Alternativa con mención en Agricultura Ecológica (Maestría en), https://www.una.ac.cr/index.php/m-oferta-academica/agricultura-alternativa-con-mencion-en-agricultura-ecologica-maestria-en, accessed 30.06.2018.

United Nations Development Programme (2016): Human Development Data (1990-2015), http://hdr.undp.org/en/data#, accessed 28.05.2018.

Vargas Garro, A. (2018): The Agroecology Movement in Latin America. Aims, actors, structures and relation to Organic Agriculture. Stüber, San José, Costa Rica & Stuttgart, Germany.

World Bank, OECD (2016): Agriculture, value added (% of GDP), World Bank national accounts and OECD National Accounts.

9. Annex

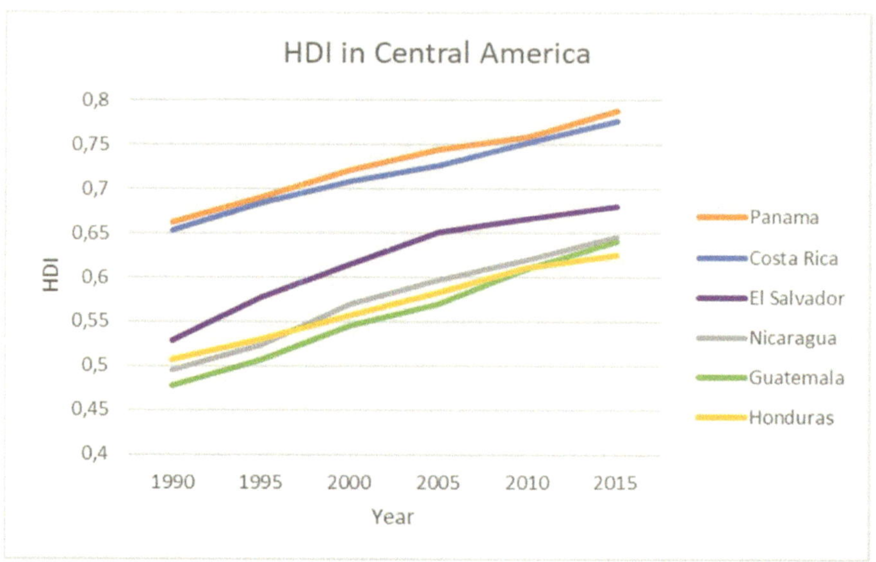

Figure 10: Human Development Index of Costa Rica (blue) and other CELAC countries in the Central American region (United Nations Development Programme 2016)

Table 2: Further information on the APPTA association and their working methods (APPTA 2017)

Explanatory video about the APPTA	https://www.youtube.com/watch?v=j8RKlJ6LlFU
Documentary by the APPTA: Agroecological production in Talamanqueña	https://www.youtube.com/watch?v=kLXTKoAM2M4
Agroforestry system of bananas, cacao and fruits, designed by the APPTA.	
Far off firm roads, the river side is an often-used transport route.